better by design

Pursuit of excellence in healthcare buildings

LONDON: HMSO

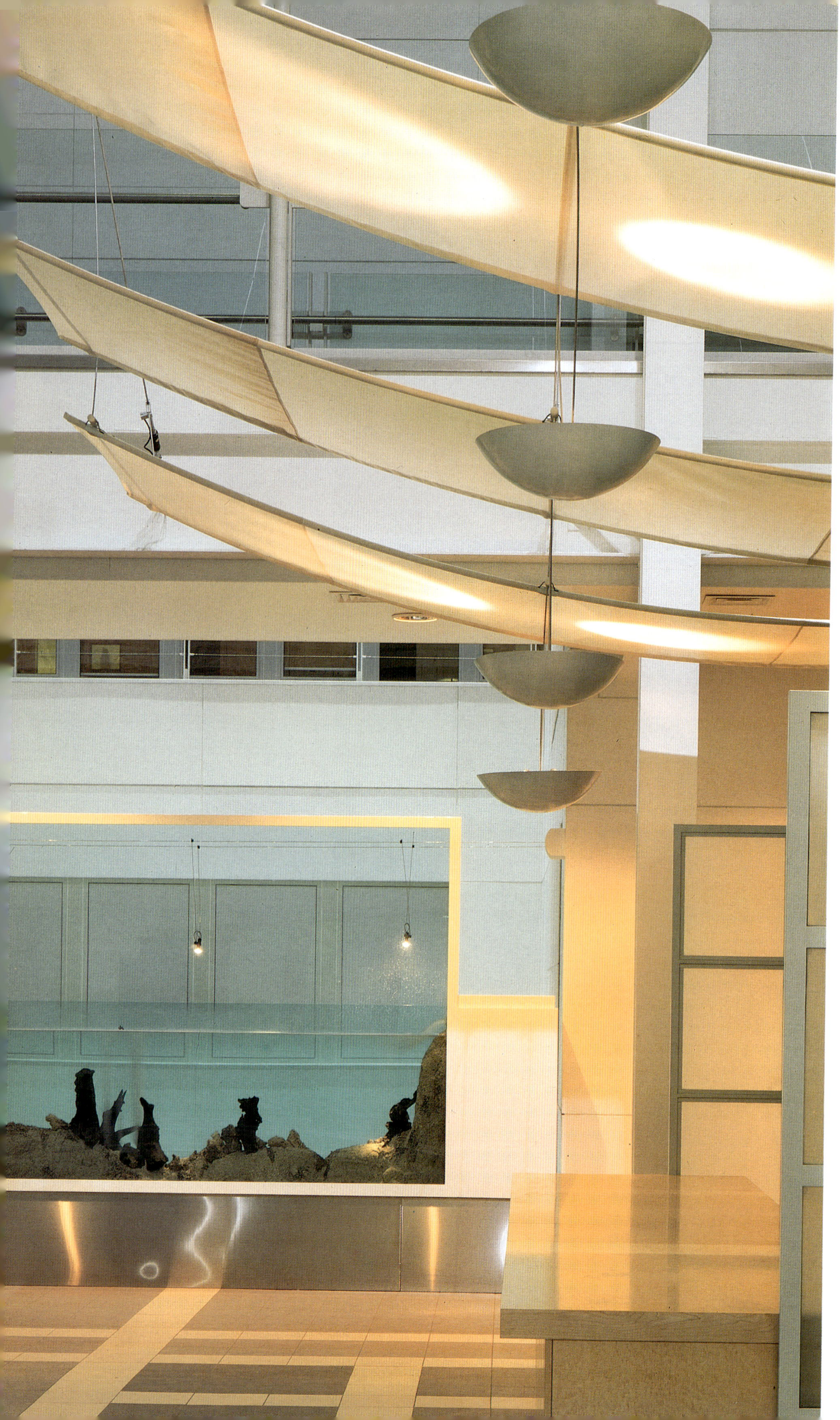

contents

About this publication 4

Foreword 5

Introduction 7

The Patron – custodian of quality 8
Setting standards
Taking the lead
Safeguarding quality

Why bother with good design? 9
Quality buildings promote health
Benevolent buildings
Working environment
Value for money
Reputation and public accountability
Public perception
Giving users what they need

Key elements of design 12
Aesthetics
Scale
The elements of surprise
The patient environment
Defining private and public space

Making the building work 14
Designing the building environment
Low energy, low maintenance
Lighting

Arts in healthcare 15

Some principles of design 16
Development control plans
Buildings as good neighbours
Integration of design ideas
Clear, logical planning
Adapting to changing needs

Commissioning a quality building 18
Organising the project
Commissioning – the key factors
Professional advice
Selection of the design team
Design competitions

Lifetime commitment to quality 20

Appendix One – Checklist of design pointers 22

Appendix Two – Consultation and the brief 24

Appendix three – A model two-stage selection process combining quality and cost 25

Appendix four – Design competition guide 27

Appendix five – Case study – a design competition . 28

Reference material 29

About NHS Estates 30

Acknowledgements 31

about *this* publication

The NHS reforms have led to greater local freedom to commission healthcare buildings of quality, durability and style, whether new or refurbished. With such freedom comes responsibility; failure to keep up quality standards will inevitably attract criticism and blame.

Quality healthcare buildings make good sense in terms of patient, staff and visitor satisfaction and in their effect on patient recovery, a factor which should be of considerable interest to purchasers.

This guide has been produced by NHS Estates to assist chairmen, chief executives and general managers in considering quality in healthcare building. It discusses the meaning of quality, and how to achieve it. Towards the end of the document, a checklist of design pointers, a model two-step procedure for combining cost and quality, a guide to design competitions, a case study and details of reference documents are also provided.

The contents of this publication are endorsed by the NHS Executive for the NHS in England.

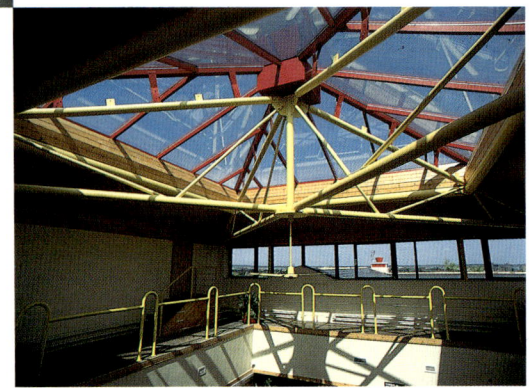

foreword

BY TOM SACKVILLE, PARLIAMENTARY UNDER SECRETARY OF STATE FOR HEALTH

"The very first condition to be sought in planning a building is that it shall be fit for its purpose. And the first architectural law is that fitness is the foundation of beauty. The hospital architect may feel assured that, only when he has planned a building which will afford the best chance of speedy recovery to sick and maimed people, will his architecture and the economy he seeks be realised."

Florence Nightingale

Ugly, impersonal buildings with forbidding interiors are hardly a prescription for recovery. Yet this is often a patient's first impression of a hospital – and it can present a stark visual contrast to the high level of medical and nursing care they find within.

Much of the estate in the NHS has suffered from poor design and bad planning, including some of its Victorian heritage, often disfigured by ugly extensions and 'temporary buildings', stained concrete monoliths, and buildings that look more like factories than hospitals.

Public buildings should be acceptable, inside and out, to the public who use them. But we have been through a period, particularly in the '60s and '70s, where much was built that can only ever have appealed to the architects themselves!

I am glad to say that things are changing. The reforms have placed greater emphasis on patient needs. We now need to ensure that this emphasis comes through in the design quality of NHS buildings.

We already have some fine examples of hospital architecture, both newbuild and refurbishment. These buildings combine a high degree of user-friendliness with flair and originality.

Quality buildings need good designers, prepared if necessary to turn conventional thinking on its head. More important still is the need for a client who will assume the role of patron, dedicated to achieving architecture in the best tradition of public building. I have asked NHS Estates when considering business cases to challenge trusts to give consideration to design quality, and be specific about how they will achieve it.

The new NHS structure will help the procurers of health buildings to meet the challenge. Decision-making has been brought closer to the grass roots. Not only do trusts have a greater responsibility towards the communities served, but there is also greater freedom than ever before to influence the quality of their estate.

Here, then, is the opportunity for chairmen to communicate a clear visual message about the services they provide and the reputation they wish to enjoy, now and in the future.

Chairmen themselves need to be closely involved at every stage. Decisions taken now affect generations to come.

TOM SACKVILLE

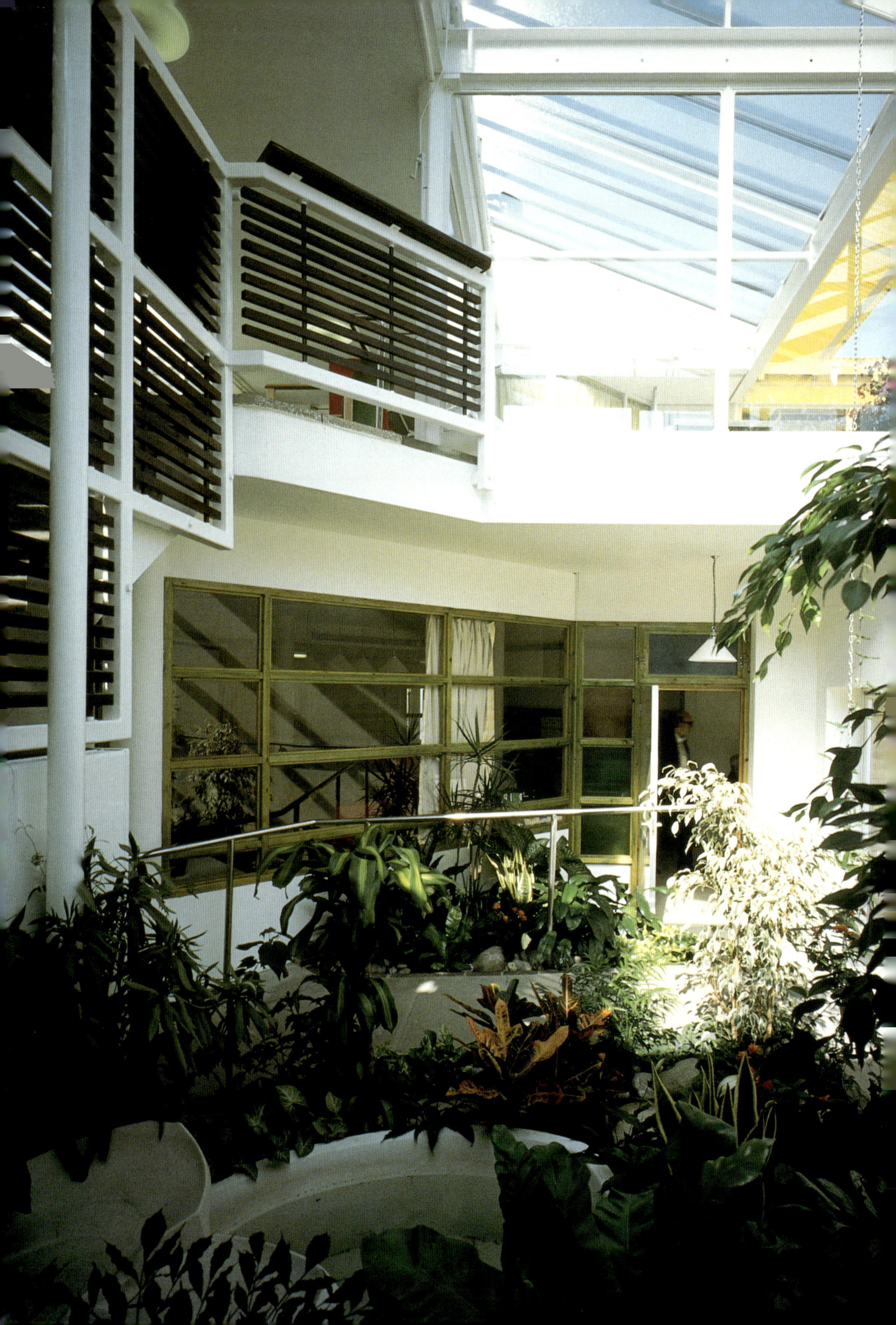

introduction

"Good design may not cost less but it need not cost more. Successful developers over the years have recognised that there is a correlation between quality and long-term value but this is always a fine balance."
Lord Sainsbury

As a major commissioner of new buildings, the NHS should be amongst the leaders in setting exemplary standards of design in public architecture. The visual impact of a healthcare building can also bring significant benefits to the sick and to those who care for them.

An attractive hospital need not cost more to build. If quality is seen as integral, not incidental, to the design and construction process, the probability of procuring fine, economical buildings which are flexible and tailored to the needs of present and future users is increased. It represents good value for money.

Design and cost should be considered together. Designers need thinking time. Realistic time and cost invested in design, a small fraction of the life cycle costs, can pay enormous dividends both in the useful life of the building and in savings on construction and long-term revenue costs.

Design quality owes as much to the vision of the client as it does to that of the design team. A combination of top-level commitment, co-operative teamwork, local involvement and the creative interplay of ideas is a fundamental requirement if the built environment of the NHS is to be improved.

WELL-DESIGNED BUILDINGS:

FUNCTION WELL;

LOOK ATTRACTIVE;

IMPROVE THE LOCALITY.

HELP PATIENTS TO RECOVER MORE QUICKLY;

HELP STAFF TO WORK BETTER;

REDUCE LONG-TERM RUNNING COSTS;

IMPROVE THE IMAGE OF BOTH THE TRUST AND THE NHS.

the patron - custodian of quality

"Quality is a state of mind, not an optional extra. It cannot be bolted on. The lead must come from a strong and committed client and the pursuit of quality must form every strand of the process."
Vincent Wang, Design Director,
Stanhope Properties plc

SETTING STANDARDS

Chairmen and chief executives hold in trust an obligation to create fine public buildings and they will be judged by the public on their success or failure in this. The health sector has the largest public capital building programme in the UK and the way the estate is handled has great impact on the environment. As patrons, chairmen and chief executives have the key role in ensuring that architectural quality is high on the list of building requirements and that the vision and values of the trust are communicated in the design.

The Patients' Charter already includes references to the estate, in terms of privacy, access and signposting. Chairmen have the scope to widen its remit locally to incorporate commitments in design terms.

TAKING THE LEAD

Commitment to quality has to be driven from the top. Standards should be set and the procedures for maintaining them communicated to everyone.

A clear statement of the brief's design objectives will avoid arguments about quality standards at a later stage which waste time and money.

The appointment of the right architect is vital. Design, however, is a team effort involving creative interaction between architect, services and structural engineers, interior designers, landscape architects and sometimes contractors.

SAFEGUARDING QUALITY

Managers and their agents and consultants should be clear from the outset about the level of quality required and be free to manage the project in the best interests of design. Safeguarding standards should then be a matter of measuring results against the criteria in the brief. A firm overview from the chairman at agreed milestones will keep the concept firmly in the hands of the commissioner and ensure that the project is kept on track.

why bother *with* good design?

> "*Mammoth hospitals, built like dreary office blocks on a devastatingly functional basis, depress the spirits, however good the care is*".
> HRH The Prince of Wales

QUALITY BUILDINGS PROMOTE HEALTH

The importance of environment to the way we live should not be underestimated. Bleak, ugly buildings depress people and attract vandals. Pleasant surroundings, on the other hand, stimulate a positive response. A holistic approach, where patients feel good about their environment as well as their treatment, will help to reduce stress and to speed recovery.

BENEVOLENT BUILDINGS

It is important that the individual is not sacrificed to the institution. Design which emphasises the maintenance of health rather than the management of disease gives a higher priority to human scale. Patients need friendly, visually attractive and non-stressful environments. A well-designed building helps people to retain their sense of identity and dignity.

WORKING ENVIRONMENT

The NHS is the largest single employer in the country and more than 80% of its running cost is spent on staff. Trusts have a responsibility towards their workforce and should aim to get the best from them.

The quality of the working environment is an indicator of the regard in which staff are held. The quality of a building's layout, safety, security arrangements and amenities is a significant factor in preventing absenteeism and sickness. Pride in a well-designed building does wonders for its staff relations, with tangible benefits for future retention and recruitment.

WHY BOTHER WITH GOOD DESIGN?

"Good design need not be expensive. Northcliffe has commissioned several buildings in recent years and an analysis of costs, updated for inflation and adjusted for site factors, indicated that good design of itself was not a prime determinant of building cost."
Christopher Carter, Director and General Manager,
Northcliffe Newspapers

VALUE FOR MONEY

We are already paying the price of poor quality design: buildings designed just twenty years ago are incurring disproportionately high running costs and maintenance bills – with all the disruption that entails – and adapt with difficulty to rapidly changing needs.

Trusts have to balance the demands of patients and staff for better quality design against the resources available. Buildings need to be efficient, economical, long-lasting and flexible. They should also be environmentally friendly.

A well-designed building represents value for money and can have enormous long-term advantage in fast payback of capital costs.

REPUTATION AND PUBLIC ACCOUNTABILITY

Quality can enhance reputation, giving an added advantage in the competition for resources. The blame for poor buildings will increasingly rest on the client as well as designers and planners:
- patients are voicing their case more forcefully, supported by the Patients' Charter and the Health of the Nation initiative;
- NHS trusts are more publicly accountable for the quality of their estates;
- interest in public architecture is increasing;
- the NHS Executive requires specific consideration of design quality to be part of the Business Case.

PUBLIC PERCEPTION

The way the public perceives hospitals and community units is important. The individuality and ideals of each trust are expressed through the style and quality of its buildings:
- do they enhance the built environment?
- are they seen as significant landmarks or a blot on the landscape?
- are they architectural good neighbours, respecting the quality of the surrounding environment or even improving it?
- are people proud of the way their health buildings look?

WHY BOTHER WITH GOOD DESIGN?

"Buildings for delivering care should themselves display care and sensitivity in every aspect of their design."
The Welsh Health Common Services Authority

Giving users what they need

Each site demands its own solutions, but the achievement of quality depends upon some key elements:
- a strong commitment to design excellence at every stage of the project;
- buildings tailored to local needs, with a high degree of user involvement;
- a successful balance between the expectations of both client and users. Whether a building pleases its customers is a major indicator of its success. This is difficult to define without clear criteria. The user benchmarks below may assist in shaping the brief and monitoring the design process.

Healthcare buildings should:

For patients and staff:

- GIVE A SENSE OF OWNERSHIP;
- BE WARM AND FRIENDLY;
- BE STRESS-REDUCING;
- BE LIFE-ENHANCING;
- PRODUCE A POSITIVE RESPONSE;

For the trust as owner:

- BE EFFICIENT IN OPERATION;
- BE ECONOMICAL IN THE USE OF RESOURCES;
- BE FLEXIBLE;
- BE DURABLE;
- BE ENVIRONMENTALLY FRIENDLY;
- SATISFY THE LOCAL PLANNING AUTHORITY;
- REFLECT THE VISION AND VALUES OF THE TRUST;

For the public:

- ENRICH THE ENVIRONMENT;
- FOSTER CIVIC PRIDE;
- BE A GOOD NEIGHBOUR;
- SIT WELL IN THEIR SURROUNDINGS;
- DEMONSTRATE INTENT;
- ENHANCE REPUTATION.

key elements of design

"A bicycle shed is a building; Lincoln Cathedral is a piece of architecture. Nearly everything that encloses space on a scale sufficient for a human being to move in is a building; the term 'architecture' applies only to buildings designed with a view to aesthetic appeal."
Sir Nikolaus Pevsner, architect

Aesthetics

Beauty is in the eye of the beholder, and therefore the patient's view of a hospital may differ from the vision of the designer. Both are subjective, but good buildings will have the visual power to uplift. This quality of inspiration, unquantifiable in financial terms, is of great value in an environment which concentrates its efforts on healing the body but may overlook the need for spiritual and emotional solace.

Design flair arises from imagination and a creative interplay between client and design team. Architects and designers respond to inspiration. Yet design flair as well as function can evolve out of practical considerations. Aesthetics and economics are indivisible, not mutually exclusive. They both derive from a preoccupation with fitness for purpose.

Domestic scale

Patients can be reassured by creating a sense of familiarity through people-orientated rooms and by softening hard, clinical contours with natural materials and lighting, soft furnishings, gardens and courtyards.

The element of surprise

Surprise and excitement handled well can be invigorating. The unexpected can be used to good effect and in a functional building owes much to context. Space is at a premium, yet space and volume can still be used dramatically while remaining within cost if concentrated on key areas such as entrances, corridors, waiting and rest areas. The dramatic effect of unusual materials, unexpected vistas or changes of level can create a jolt of wonder and a sense of great enjoyment.

KEY ELEMENTS OF DESIGN

The patient environment

Good hospitals have a feeling of harmony. The appropriateness of materials, colour schemes, interior decoration and artwork, water and landscaping can contribute to a sense of calm. The use of natural light and ventilation gives a link with life outside, a degree of individual environmental control and is economically preferable to remote controls.

Views are important – both of external landscapes and of internal courtyards or artworks. They can appear to extend space and provide a secure connection to the outside world. Patients should not be made to feel cut off from the activity going on around them both inside and outside the building.

People respond to particulars, not plans. Detail is just as important as the grand concept and should be integrated into the overall design, the landscape and interior.

Defining private and public space

People like to own their space, whether it is the few square feet around a bed, or a public waiting area. There is a hierarchy to the spaces within a health building: it is important to establish their relative value and to define both public and private space. Long corridors, for example, can be punctuated and enlivened; they are not just a means of moving people and goods from one place to another, but are important areas of common ground, necessary for private encounters and conversations.

See Appendix One for a checklist of design pointers.

13

making the *building* work

"The creative art of directing the forces of nature."
Professor Sir Edmund Happold, Engineer

Technical services are always an integral part of good architecture. In the right context, inspired structural, mechanical and electrical services solutions are frequently the key to a creative building.

Designing the building environment

Money and design time spent on the right building management systems pay off handsomely in the long run. Engineering services, from central heating plant to nurse-call systems, are crucial to ensuring the comfort and safety of patients and staff.

Low energy, low maintenance

Hospitals are potentially high consumers of energy, but there are a number of design factors which can improve a building's energy performance, from insulation and ventilation to thermostatic central heating and hot and cold water systems. Recommendations in 'The Low Energy Hospital Study' (published by NHS Estates) aim to reduce energy costs by up to 50%.

Lighting

The quality of light in a hospital building is important to the well-being of patients who spend most of their day confined to a ward. Artificial lighting should be as natural as possible in patient areas, avoiding eye strain and glare while enhancing the interior design. In clinical areas it must satisfy different, but equally important, criteria. In both situations lighting should be fully integrated with natural light sources and be easily accessible for maintenance.

arts *in* healthcare

"That the arts can be therapeutic is not of course a new idea. But it is an idea whose time has come."
Lord Attenborough
President, Arts for Health

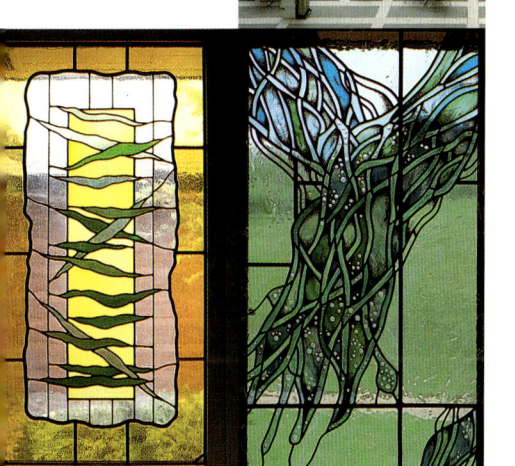

Healthcare buildings can play a real part in the community through specially commissioned art. Work by local artists and craftspeople is important in erasing the institutional feel, providing patients with creative reminders of their locality and endowing the building with cultural landmarks.

There are many forms of creative expression, from painting, craftworks and photography to murals and sculpture. Providing spaces for the performing arts is becoming increasingly popular in hospitals. Both amateurs and professionals may be involved but skill and experience are vital in ensuring high standards and suitability for purpose.

Arts Council and local funding may be available, but resources may also need to be earmarked, and fund-raising campaigns launched by targeting charities, health endowment and trust funds, businesses, arts organisations and the local community.

Trusts should consider establishing a steering committee for incorporating the arts into the new building. It should comprise a good cross-section of people, including a senior manager, all of whom are positive about the project. One person should be responsible for drawing up a coordinated programme, with the help of professional advisers, and for involving the hospital and local community as much as possible.

some principles of design

"First impressions tend to stick. Negative first impressions will perhaps damage patient confidence. Good experience will help the caring process and make the job of other colleagues down the line easier and more effective. How often we see quality clinical treatment being compromised by drab inappropriate physical surroundings."
Bill Murray, Chief Executive,
South Tees Acute Hospitals NHS Trust

Good design cannot be tacked on at the end of a project to dress up the externals. Well-planned buildings place the needs of users uppermost and are designed from the inside outwards. The design concept provides an underlying philosophy which informs all decisions and ensures that they are appropriate and consistent.

DEVELOPMENT CONTROL PLANS

All healthcare sites need masterplans. The right design input at an early stage will ensure that new buildings not only harmonise with their existing surroundings but communicate logically within the hospital 'townscape' and avoid piecemeal development in the future.

BUILDINGS AS GOOD NEIGHBOURS

Trusts may wish buildings to be prominent local landmarks or, in the case of hospices or long-stay hostels, less conspicuous. The siting of a building is important and has a strong bearing on design. Considerations include the visual impact of the building in relation to its neighbours: its storey height, scale, fabric and character. For example, does it make use of local views?

Approaches to a hospital and the entrances can convey an important message to staff, public, visitors and patients, by using space and colour to make a point of impact and to create a particular type of atmosphere.

Access and circulation should take account of local concerns. The provision of sufficient car parking will avoid overspill parking in residential areas, and careful landscape design can mitigate the impact of large areas of cars.

Buildings should be integrated with the site and its contours, giving the best possible orientation from within and without. The surrounding space is just as important as the building. Both can draw inspiration from the character of the locality, through choice of materials and colours, landscaping and planting.

Environmental impact should be addressed at the earliest feasible stage. Early consultation will establish town planners' feelings about the site, highlighting problems such as conservation constraints. A constructive dialogue with planners at the initial stage of a project is more likely to result in satisfactory planning approval. Failure to consult early enough can lead to confrontation and serious cost and timetable implications.

SOME PRINCIPLES OF DESIGN

INTEGRATION OF DESIGN IDEAS

All aspects of design should be considered together at the initial planning stage. Close, early teamwork between architects, surveyors, engineers, landscape and interior designers and artists, together with sufficient time for co-ordination, can contribute to both the functional and aesthetic success of the scheme.

CLEAR, LOGICAL PLANNING

Health buildings should be intelligible, yet many are like a maze and people find them hard to comprehend and negotiate.

A building's clarity reflects its priorities. For example, the desire to close the gap between hospital and community may lead to the design of a shopping mall as a focal point; elsewhere staff facilities may be higher on the agenda. These levels of priority can be interpreted through good design, with time and money concentrated on selected areas, for maximum effect.

Clarity is vital:
- entrances should be obvious;
- relationships between different departments should be logical;
- back-up facilities should be accessible.

Signposting should be well-positioned, with clear and easy-to-read graphics. Co-ordinated with interior design, these can help to create a corporate trust identity which runs throughout the building.

ADAPTING TO CHANGING NEEDS

The health buildings of today need to be flexible enough to adapt to changing needs. Besides planning for possible new extensions, existing spaces can be designed for use in a number of different ways without any need for physical alteration. For instance, a day room could also be the focal area for community activities and bazaars. Rooms can also serve several different users without becoming the sole property of any one. This non-territorial use of space is particularly useful in out-patient departments and doctors' surgeries.

commissioning *a quality* building

"*Advice should concentrate on broad matters of scale, density, height, massing, layout, landscape and access. It should focus on encouraging good design, rather than stifling experiment, originality or initiative.*"
Department of the Environment

"*There must always be a competent, informed client.*"
Sir Duncan Nichol,
former Chief Executive
NHS Management Executive

ORGANISING THE PROJECT

A well-organised project, with strong leadership, is vital to the success of any building. The first step is the trust's strategic direction which should provide, in clear and unambiguous terms, a statement of the vision and values of the trust together with an outline service profile/ philosophy. Both the project director and project manager need to establish clear lines of communication and accountability. Full guidelines are available in the Project Organisation handbook in the Capital Investment Manual. High quality design advice is essential. A board member with a special remit to champion quality and environmental issues would help to ensure the right level of advice, at the right time, from health planners, engineers, quantity surveyors and architects. Their proposals for the disposal of land and buildings may signal the downsizing of requirements, or refurbishment rather than newbuild, resulting in major capital and revenue savings.

COMMISSIONING – THE KEY FACTORS

Good briefing is the cornerstone of good design. It is the result of close teamwork between client and design team, with a strong input from the users. This is where the aims, requirements and cost thresholds of the client mesh with the creative response of the design team, setting the parameters for the project and taking the planning and site considerations into account. Therefore:

- define a realistic timescale. This does not imply excessively long programmes, but sensible planning, with a realistic period for each stage of the project;
- set a realistic budget, commissioning a feasibility study if necessary. Late cost-cutting has an adverse effect on quality;
- appoint professional consultants with a proven track record, known quality standards and the resources to carry out the commission;
- select contractors according to experience, status, stability, location and recent performance;
- set up clear lines of communication between project director, project manager and design team;
- establish a full understanding of the design process, cost evaluation, reporting and monitoring as the scheme develops;
- monitor on-site the quality of work, progress to programme and compliance with specifications.

See Appendix Two for further guidance on the development of the brief.

Professional advice

The creative use of design advisers in the procurement process can be useful in encouraging sponsors to adopt a bolder, more innovative approach to design, while providing the security of a safety net; for example:
- engaging a design adviser to input during preparation of the brief;
- using a design "champion" to assist in judging competitions and assessing submissions;
- getting a design "second opinion" when evaluating the design team's proposals (value engineering);
- using a design "partnership", where experienced health designers link skills with designers from outside this specialist field.

Selection of the design team

Quality should be built into the selection process, together with suitable safeguards. The starting point is the selection of the design team. The right design team should:
- be capable of delivering a high-quality solution;
- be given sufficient time to design in depth;
- be given sufficient fee income to enable it to allocate the right level of resources for each project.

The level of design quality required, the particular characteristics of the project and the prevailing market conditions will determine the appropriate process for the selection of the design team and, ultimately, the building procurement method.

See Appendix Three for a model two-stage selection process combining quality and cost.

Design competitions

Pursuing quality means allowing for innovative ideas and creative ways of solving conventional problems. Well run design competitions, carefully tailored to the type of building and its objectives, are an effective way of achieving a high level of design quality and stimulating fresh talent. They encourage the involvement of users, staff and the public at large and foster a greater sense of pride and ownership. Normally, a minimum of three firms should be invited to compete, although this may be modified by EC requirements.

See Appendices Four and Five for further guidance on running a design competition.

lifetime commitment to quality

"Architecture is also about the spiritual needs of people as well as their material needs. It has as much to do with optimism, joy and reassurance; of order in a disordered world; of privacy in the midst of many; of space in a crowded site; of light on a dull day. It is about quality."

Sir Norman Foster,
RIBA Gold Medal winner

Design integrity extends beyond the conclusion of the building contract. Many designs have been compromised by unsightly additions built on to courtyards or flat roofs. Safeguards need to be made if the building's future quality is to be preserved. Designers will give guidelines on maintaining the interiors and style of their buildings. Staff can also be involved in looking after their new workplace.

The completion of a building should also be followed up by an evaluation of its design effectiveness. Independent research into the measurable effects of a building's design on its users – from functional effectiveness through patient recovery rates to staff performance – could enable procurers to make future decisions with more assurance and authority.

Long life, user acceptability and buildings which stand the test of time in the public's affections are ideals which this document has endeavoured to promote. Getting it right is far from easy, but following the pointers in this guide should go a long way towards helping trusts to produce buildings which win the approval of this and future generations.

APPENDICES

appendices

APPENDIX ONE

CHECKLIST of design pointers

First impressions

Is access safe and easy, by private and public transport and on foot?

Are entrances and car parks well signposted?

Do access and parking create problems for the neighbourhood?

Do car parking facilities intrude on the landscape?

Are car parks and entrances easy to find and located close to the building?

Are entrance approaches and areas welcoming?

Does the building overwhelm its site or its users?

Does its external imagery relate to its purpose?

Is the scale of the building in keeping with its neighbours?

Is the character of the building's position reflected in its visual style?

Do materials and colours harmonise with the context?

Is exterior detailing consistent?

Does the building have clean lines?

Do service details, for example boilers, flues, lifts, detract from or contribute to the design?

Are internal circulation routes clear and convenient?

Is signposting visible and legible? Does it reflect the house style?

User-friendly features

Are departments clearly orientated in relation to the entrance, easily accessible and with minimum walking distance?

Are the needs of the disabled taken fully into account?

Is the internal streetscape reassuring rather than clinical?

Are blandness and monotony avoided, and space used to maximum effect, with plenty of variety horizontally and vertically?

Is care given to the provision of adequate private and personal space?

Can equipment be hidden from view where appropriate (while remaining secure and accessible for maintenance and repair)?

Is there provision for the work of artists and craftspeople?

Is natural light allowed into the building via windows, courtyards and roof lighting (but see also "Building efficiency" below)?

Do patient areas and staff facilities look out on to views?

Are colours warm, comfortable and harmonious? Are room schemes co-ordinated to provide continuity or variety where required?

Do furnishings take account of the visual and aural comfort of users, avoiding disturbing colour combinations, hard, noisy finishes, and using carpets where possible?

Do furniture and furnishings co-ordinate with the overall design?

Is detailing carefully thought through and consistent?

APPENDIX ONE

IS JOINERY CAREFULLY DESIGNED, PARTICULARLY IN AREAS OF HIGH CIRCULATION?

IS LIGHTING SUBTLE, LOW-GLARE AND UNOBTRUSIVE IN PATIENT AREAS (BUT CONSISTENT WITH CLINICAL REQUIREMENTS)?

THE SITE AND LANDSCAPING

IS THE BUILDING INTEGRATED WITH THE SITES CONTOURS?

IS SECURITY BUILT-IN?

DOES ORIENTATION TAKE BEST ADVANTAGE OF THE OUTLOOK (SEE ALSO "BUILDING EFFICIENCY" BELOW)?

DOES IT PROVIDE GOOD VIEWS FROM EVERY ANGLE?

DOES IT INTRUDE UPON OR BLOCK LOCAL VIEWS OR LANDMARKS?

ARE THE SPACES AROUND THE BUILDING SEEN IN RELATION TO THE BUILDING ITSELF?

DOES THE RELATIONSHIP BETWEEN INTERNAL AND EXTERNAL VIEWS AND SPACES HAVE BENEFITS FOR THOSE INSIDE LOOKING OUT AND THOSE FROM THE OUTSIDE LOOKING IN?

CAN EXISTING TREES AND PLANTING BE INTEGRATED INTO THE DESIGN?

HAVE EXISTING RIGHTS OF WAY, PRESERVATION ORDERS ETC BEEN RESPECTED?

ARE THE CHARACTERISTICS OF THE LOCAL LANDSCAPE REFLECTED IN THE CHOICE OF PLANTS AND MATERIALS?

DOES LANDSCAPING GIVE PATIENTS A VARIED AND ATTRACTIVE OUTLOOK ALL YEAR ROUND?

IS PLANTING PLANNED FOR MINIMUM MAINTENANCE?

DO GARDEN AREAS AND COURTYARDS NEED TO BE DESIGNED NOT ONLY AS ATTRACTIVE EXTERNAL SPACES BUT ALSO AS A CONTROLLED ENVIRONMENT FOR PATIENTS' RECUPERATION AND THERAPY?

BUILDING EFFICIENCY

ARE SERVICE SYSTEMS WELL INTEGRATED WITH THE STRUCTURE AND INTERNAL LAYOUT OF THE BUILDING?

ARE HEATING AND VENTILATION COMPATIBLE WITH COMFORT?

ARE SAFETY AND SECURITY STANDARDS COMPLIED WITH?

IS THE BUILDING EASY AND ECONOMICAL TO MAINTAIN?

ARE ENERGY CONSERVATION MEASURES, INCLUDING INSULATION, FUEL, SOLAR GAIN, DOUBLE GLAZING (WHERE APPROPRIATE) AND OTHER FORMS OF DRAUGHT EXCLUSION TAKEN INTO ACCOUNT?

HAS THE ORIENTATION OF THE BUILDING BEEN FULLY TAKEN INTO ACCOUNT WHEN PLANNING SERVICE SYSTEMS, FOR EXAMPLE SOLAR GAIN?

HAVE FABRIC AND MATERIALS BEEN CHOSEN FOR THEIR DURABILITY?

ARE INTERIORS EASY TO CLEAN, ACHIEVING A BALANCE BETWEEN WASHABILITY AND GOOD SOUND ABSORBENCY?

IS THE BUILDING CAPABLE OF ADAPTATION TO FUTURE NEEDS, THROUGH CHANGE OF USE OR PHASED EXTENSION?

APPENDIX TWO

CONSULTATION and the brief

THE BRIEF

A well thought-through, comprehensive brief which looks at every angle will help prevent mistakes and omissions filtering through the commissioning process to cause major problems on site.

KEY ELEMENTS OF THE BRIEF

The key elements of the brief are:

- A strong statement linked to the aims and values of the trust, describing the key objectives, in line with the business case and with a clear set of quality benchmarks;

- A description of the intended impact of the scheme on the local environment;

- Whole-hospital and departmental operational policies, with projections about the provision of existing and future services;

- A full functional content for the building with schedules of accommodation for each department, describing each space with its area (M^2). Target circulation areas within departments and communications between departments, including plant space, should be provided.

CONSULTATION WITH USERS

NHS guidelines (for example, Health Building Notes and Health Technical Memoranda) should be tested against specific user needs via consultation meetings. A full range of users should be consulted, including relevant staff, patients and those with disabilities.

CONSULTATION WITH PLANNING AUTHORITIES

The abolition of Crown immunity for healthcare buildings has brought the importance of town planning to the fore. Early comment on the urban planning implications is essential.

CONSULTATION WITH INTEREST GROUPS

National and local organisations can be a source of help and advice, such as English Heritage, the Royal Fine Art Commission, the Department of the Environment Conservation Unit, and Listed Buildings Panels (which also coordinate interest groups such as the Georgian Group). Local historic, conservation and environmental groups can also help to formulate site and energy strategies.

APPENDIX THREE

A MODEL TWO-STAGE SELECTION PROCESS combining quality and cost

This process may be used by clients who are wary of the potentially damaging effects on quality of unqualified fee bidding, but who do not have the time and/or inclination to hold a design competition.

Step One

The client (assisted by the consultant and project manager, if applicable) compiles a list of suitable design consultants. The list should be between four and eight firms from each discipline. If value dictates, the project must be advertised in the EC Journal and a minimum of five firms should be included on this list (for each discipline).

Step Two

Each listed firm is sent the project brief and asked to submit a proposal, specifically tailored to the project, containing:

- a brief description of the firm, highlighting relevant experience;

- a description of the roles of the key personnel proposed, plus their CVs;

- a project methodology, setting out how the project would be designed and managed;

- answers to key relevant questions relating specifically to the project, for example advice on suitable building contract types.

The project proposal should not exceed ten sides of A4 to ensure that responses are focused and that the assessment procedure is simplified.

Step Three

The client and project manager evaluate the project proposals against a predetermined set of criteria. They then select a shortlist of firms from each discipline and invite them to attend an interview with relevant members of the client body.

The invitation for interview should make it clear what specific issues will be addressed, the length of the interview and who is expected to attend. The intended lead professional directly responsible for the project and the project professional (for example the project engineer) should be invited.

Step Four

The interview panel must be consistent for all interviews.

Interview assessment must follow immediately and evaluation forms must be completed before interviewing the next consultant.

An interview assessment proforma states the key criteria to be addressed. A weighting is applied for each criterion and each member of the interview panel assesses the performance of the consultants on a scale of, say, 1 to 10. The score is then multiplied by each weighting to give a total score. At the end of all interviews each consultant's performance is again reviewed and scores adjusted.

The firms with the highest scores are then invited to tender, normally a minimum of four (minimum of five if the EC Directive applies). They should be asked to submit a resources chart showing in detail the proposed personnel and the number of hours allocated to each stage of the project.

APPENDIX THREE

STEP FIVE

THE FINAL SELECTION IS BASED UPON AN EVALUATION OF FEE OFFERS INVOLVING QUALITATIVE AND FINANCIAL CRITERIA TO DETERMINE VALUE FOR MONEY. THE QUALITATIVE CRITERIA SHOULD INCLUDE:
- EVALUATION OF PROJECT PROPOSALS
- PERFORMANCE AT INTERVIEW
- EVALUATION OF RESOURCES OFFERED.

SEE ALSO: THE ASSOCIATION OF CONSULTING ENGINEERS "BALANCING QUALITY AND PRICE", VALUE ASSESSMENT AND THE SELECTION OF CONSULTING ENGINEERS.

APPENDIX FOUR

DESIGN COMPETITION
guide

COMPETITION GUIDE

The Royal Institute of British Architects (RIBA) Competition Office (8 Woodhouse Square, Leeds LA3 1AD. Telephone 0532 341335) offers advice and will administer competitions for clients.

FEE OPTIONS

Design competition fees
Non-rewarded design competitions are rightly seen by designers as an exploitation of their concept ideas. It is recommended that all competitors are paid a fee or premium for their competition entry. One method is to calculate the recommended RIBA fee for a sketch design and divide this among the competitors. The winning competitor is paid from then on. The winner should also be paid a prize that can be set against future design fees.

Project fees
Fees relating to design competition-generated commissions can be established in one of two ways:
1) Competitors are asked to submit fee proposals alongside their design submissions;
2) Once the winning design is selected, the client negotiates the fee with the winner.

STEP ONE – SELECT COMPETITORS

Competitors can be selected using a number of criteria including their design ability and appropriate track record. If the EC Directive applies, advertise the intention to hold a design competition in the EC Journal, asking interested participants to inform the RIBA Competition Office.

STEP TWO – APPOINT ASSESSMENT PANEL

Appoint an appropriate panel of competition assessors, for example, two high grade architects/designers with appropriate experience, the chairman, the project director, the estates manager and one or two key user representatives.

STEP THREE – SELECT SHORTLIST

Select a shortlist of not more than six competitors in collaboration with the assessors and the RIBA.

STEP FOUR – DRAW UP BRIEF

Draw up a detailed brief for the project, taking full account of qualitative issues and functional requirements.

STEP FIVE – SEND TERMS

Send the shortlisted competitors the competition terms, the brief and an invitation to attend a formal question and answer session with the assessors. If the scheme is subject to the European Community Services Directive, all interested competitors should be sent the rules.

STEP SIX – ANSWER QUESTIONS

Receive and answer all relevant questions from competitors at the briefing session. Record these and circulate them to all competitors. No other contact between competitors and assessors/client should occur except to clarify essential points. These should then be sent to all competitors.

STEP SEVEN – ESTABLISH ASSESSMENT METHODOLOGY

With the help of the RIBA and the external assessors, establish a structured assessment methodology for judging the design submissions, giving appropriate weighting to the priority issues.

STEP EIGHT – ASSESS ENTRIES

Receive the entries at the appointed time, ensure there are no distinguishing features whereby the identity of the competitor can be detected, and make the assessment in accordance with the criteria established.

APPENDIX FIVE

CASE STUDY –
a design competition

South Downs Health NHS Trust chose a promoter choice route for a small £600,000 clinic, demonstrating that the competition method need not apply solely to major projects. Six widely differing practices were shortlisted, some private, some public, some local, some regional in scope. Equally important, some practices were chosen because they had no previous experience of the design of health buildings and hopefully would bring with them new insights and interpretations.

The four assessors comprised an architect, a local architect who understood the area, and two involved Trust managers. A small technical panel of community nurses, health visitors, GPs and local councillors also made their comments at an exhibition of the entries on the housing estate where the new clinic was to be constructed. These comments were passed to the assessors.

Each competitor was paid a premium to enter, with the winners prize set against the project fees. The final fees were negotiated with the winning design team. It was the promoter's intention to make the winning team study every aspect of the future building performance in great detail and thereby achieve far greater savings long-term than could be achieved by paring fees at the front end of the project. The six designs were reduced to two, from which Trust board members themselves made the final decision. A major advantage of this method was the enthusiasm and involvement of staff at all levels, who felt that not only were their ideas listened to, but they actually influenced the choice of design solutions.

This method was so successful that it has been used again by the Trust for a substantially larger community hospital.

reference *material*

DOCUMENTS

CONCODE – Gives guidance and sets out mandatory contracting requirements for NHS capital building projects.

ENCODE 1 AND 2 – Give guidance and practical advice on energy efficiency in healthcare buildings.

A strategic guide to environment policy (1992).
A strategic guide to combined heat and power (1993).
A strategic guide to water and sewerage policy (1993).
A strategic guide to clinical waste management (1994).
A strategic guide to energy policy (1994).

The strategic guides are published by and available from NHS Estates.

ORGANISATIONS

Arts for Health

British Healthcare Arts

Chartered Institution of Building Services Engineers (CIBSE)

Institution of Electrical Engineering (IEE)

Institute of Landscape Architects (ILA)

Institution of Mechanical Engineering (IMechE)

Royal Institute of British Architects (RIBA)

Royal Institution of Chartered Surveyors (RICS)

about NHS Estates

NHS Estates is an Executive Agency of the Department of Health and is involved with all aspects of health estate management, development and maintenance. The Agency has a dynamic fund of knowledge which it has acquired during 30 years of working in the field. Using this knowledge NHS Estates has developed products which are unique in range and depth. These are described below.

NHS Estates also makes its experience available to the field through its consultancy services.

Enquiries should be addressed to: NHS Estates, 1 Trevelyan Square, Boar Lane, Leeds LS1 6AE. Tel: 0532 547000.

SOME OTHER NHS ESTATES PRODUCTS

ACTIVITY DATABASE – a computerised system for defining the activities which have to be accommodated in spaces within health buildings. *NHS Estates*

DESIGN GUIDES – complementary to Health Building Notes, Design Guides provide advice for planners and designers about subjects not appropriate to the Health Building Notes series. *HMSO*

ESTATECODE – user manual for managing a health estate. Includes property management advice and a recommended methodology for property appraisal. Provides a basis for integration of the estate into corporate business planning. *HMSO*

CONCODE – outlines proven methods of selecting contracts and commissioning consultants. Reflects official policy on contract procedures. *HMSO*

WORKS INFORMATION MANAGEMENT SYSTEM – a computerised information system for estate management tasks, enabling tangible assets to be put into the context of servicing requirements. *NHS Estates*

HEALTH BUILDING NOTES – advice for project teams procuring new buildings and adapting or extending existing buildings. *HMSO*

HEALTH GUIDANCE NOTES – an occasional series of publications which respond to changes in Department of Health policy or reflect changing NHS operational management. Each deals with a specific topic and is complementary to a related Health Technical Memorandum. *HMSO*

HEALTH TECHNICAL MEMORANDA – guidance on the design, installation and running of specialised building service systems, and on specialised building components. *HMSO*

HEALTH FACILITIES NOTES – debate current and topical issues of concern across all areas of healthcare provision. *HMSO*

ENCODE – shows how to plan and implement a policy of energy efficiency in a building. *HMSO*

FIRECODE – for policy, technical guidance and specialist aspects of fire precautions. *HMSO*

MODEL ENGINEERING SPECIFICATIONS – comprehensive advice used in briefing consultants, contractors and suppliers of healthcare engineering services to meet Departmental policy and best practice guidance.

Items noted "HMSO" can be purchased from HMSO Bookshops in London (post orders to PO Box 276, SW8 5DT), Edinburgh, Belfast, Manchester, Birmingham and Bristol or through good booksellers.

Enquiries about NHS Estates should be addressed to: NHS Estates, Marketing and Publications Unit, Department of Health, 1 Trevelyan Square, Boar Lane, Leeds LS1 6AE.

NHS ESTATES CONSULTANCY SERVICE

Designed to meet a range of needs from advice on the oversight of estates management functions to a much fuller collaboration for particularly innovative or exemplary projects.

Enquiries should be addressed to: NHS Estates Consultancy Service (address as above).

acknowledgements

ARCHITECTURE:

Abbey Hanson Rowe
Avanti Architects Limited
Ahrends Burton and Koralek
Carnell Green Nightingale
Edward Cullinan Architects
Estatecare Group, trading as part of Welsh Health Common Services Authority
Alex Gordon Partnership
Fairhursts Design Group
HLM Architects
Llewelyn-Davies
Medical Design Partnership Ltd
Pentarch Ltd
Percy Thomas Partnership (Architects) Ltd
PTP Landscape and Urban Design Ltd
Powell Moya Partnership
Sheppard Robson
Douglas Stephen and Partners

PHOTOGRAPHY:

Matthew Antrobus
Graham Challifour
Martin Charles
Anne Dick
Chris Edgcombe
D Gilbert
Ed Horwich
John Edward Linden
Matthew Weinreb
Charlotte Wood

ARTWORK:

Lee Cheong, MDP Interior Design
Allen Jones
Shona McInnes
Hilary Salisbury

BROCHURE DESIGN:

...Talk Design Communications

OTHER CONTRIBUTIONS:

Thanks to those trust chairmen, chief executives and professionals who have supported the production of this publication.

better by design

Pursuit of excellence in healthcare buildings